MY BODY AND HOW IT WORKS

Illustrated by Barbara Harmeyer

A TOM DOHERTY ASSOCIATES BOOK
NEW YORK

MY BODY AND HOW IT WORKS

A TOR Book
Published by Tom Doherty Associates, Inc.
49 West 24th Street
New York, NY 10010

ISBN: 0-812-59487-8 Can. ISBN: 0-812-59488-6

Library of Congress Catalog Card Number: 88-50337

First edition: October 1988

Printed in the United States of America

0 9 8 7 6 5 4 3 2 1

Have you ever wondered what makes your body move, why your heart beats without having to think about it, or what happens to the food you eat? Some of the answers will amaze you. In this book you'll find answers to many of the questions that kids like you often ask about the human body and how it works.

What Is My Body Made Of?

Seventy Percent of Your Body Is Water

Water is a compound made of *elements*. An element is one of the basic materials from which everything is made. There are about 100 elements. Just as letters of the alphabet combine to make words, elements combine to make *compounds*. *Hydrogen* and *oxygen* are elements. Together they form the compound known as water. Most of your body is water. The rest of your body is made of other compounds. Water helps to move materials through your body. Some of these materials dissolve in the water. Some are moved in other ways.

DISCOVERY CORNER:
Try this experiment and find out which of the test materials dissolve in water. You will need 2 glasses of water, a tablespoon of salt, and a tablespoon of cooking oil (fat). First, add the oil to one of the glasses of water and stir. Do the oil and water combine easily or do they remain separate? Now, add the salt to the second glass of water and stir. Which material dissolved? Which did not?

What Is the Smallest Part of the Body?

The Cell

Our bodies are made up of cells. Some are so tiny that more than one hundred thousand of them would fit on the head of a pin. One of the smallest is the *red blood cell*. In each fluid ounce of healthy human blood there are more than 150 billion of these tiny cells. There are many different kinds of cells in the body, and they are different sizes and shapes. Each has a special job to do. Every part of the human body — bones, skin, muscle, even hair — is made up of, or produced by, cells.

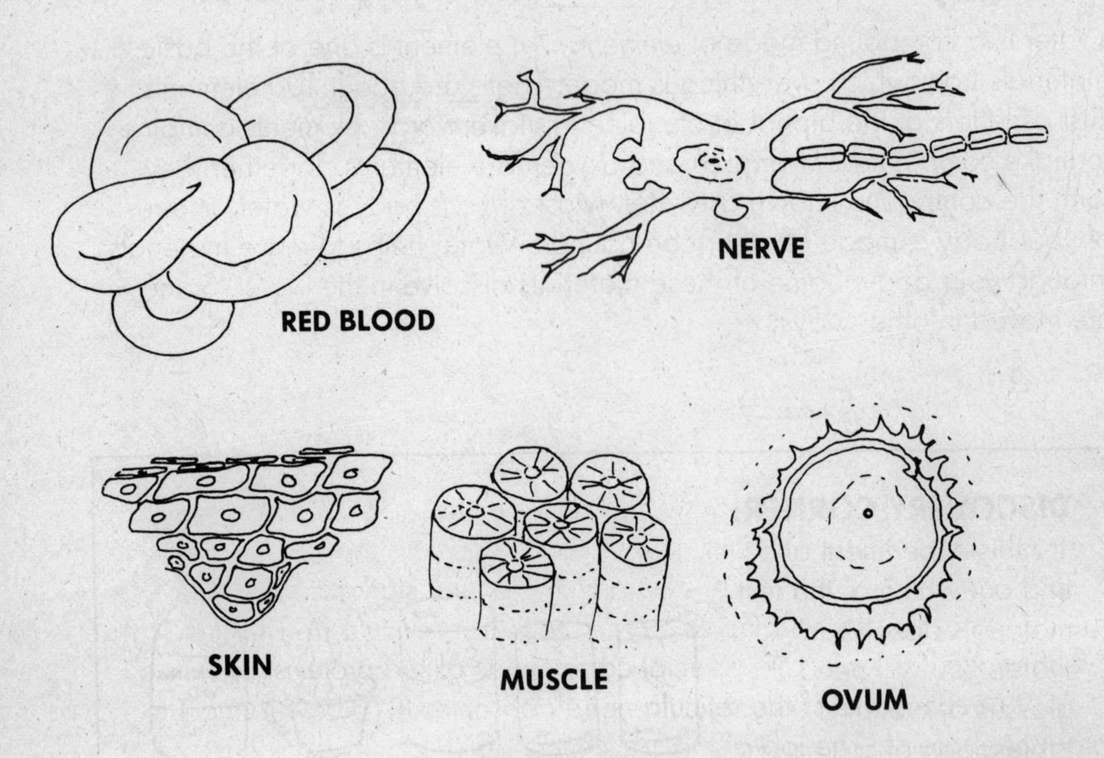

FUN FACT: There are 500 times more cells in the human body than there are known stars in our galaxy.

What Is the Inside of a Cell Like?

There Are Different Things Inside Different Types of Cells

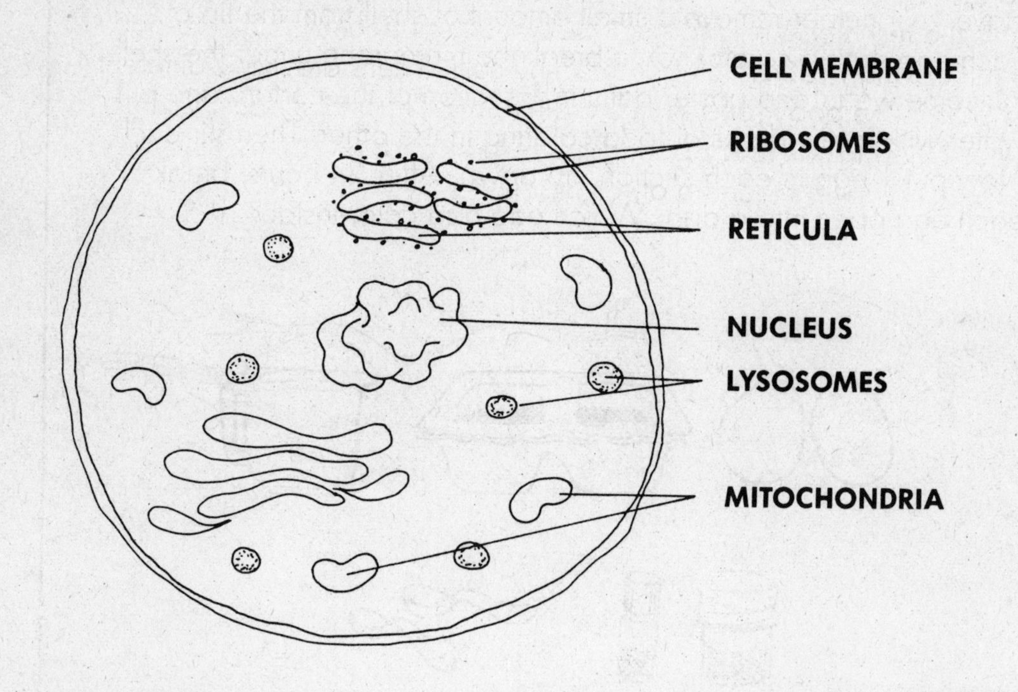

Many cells are like small cities with different parts, and each part has its own job. The *nucleus* is the command center that tells the cell what to do. *Mitochondria* (my-toe-KON-dree-a) are the power stations. They produce the chemical energy needed to run the cells. The *reticula* (re-TICK-you-la) are the factories that make special compounds called *proteins*. Just as a factory needs workers, the reticula need *ribosomes* (RYE-bow-zomes) to produce proteins. The job of the *lysosomes* (LIE-so-zomes) is to dispose of harmful particles and worn-out parts. The cell *membrane* is like a thin, protective wall of fat around the cell. It allows only certain things to pass in or out of the cell.

Not all cells are like this inside. For instance, mature red blood cells do not have all of these parts. Instead, they contain a compound called *hemoglobin* (HEE-mow-glow-bin), which carries oxygen in the blood.

DISCOVERY CORNER: Here is an experiment that shows how membranes allow only certain materials to pass. You will need 2 raw eggs, 2 sections of a plastic egg carton, blue poster paint, red food coloring, water, 2 spoons, and a little help from a grown-up. Have your helper remove a small amount of shell from the tip of each egg, being careful not to break the membrane under the shell. Put some water and poster paint in 1 section of the carton, and put water with a few drops of food coloring in the other. Then stir each. Now put 1 egg in each section, tip down. After 24 hours, break each egg open into a dish. Which egg had color inside?

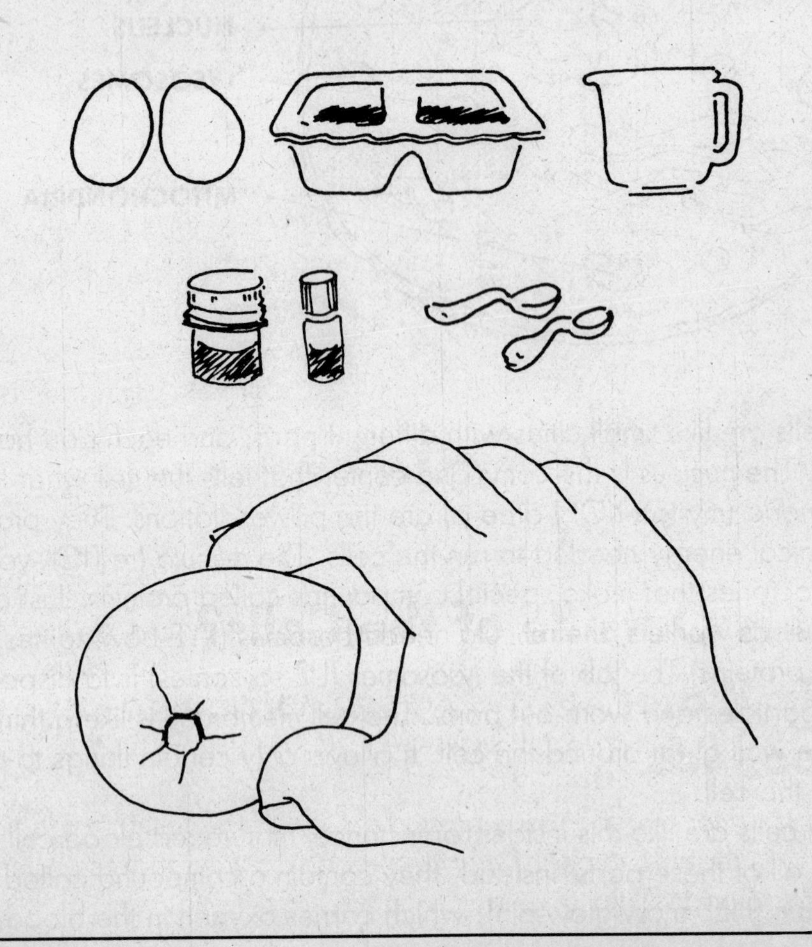

FUN FACT: Some cells of the body are large enough to see. Look through a magnifying glass at the back of your own hand. The top layer of skin cells are easy to see.

How Do Cells Know What To Do?

Cells Carry Instructions in the Nucleus

These instructions are called *genes*. They tell the cell what kind of cell it will be, which jobs it will do, and which products it will produce. Your genes come from your parents, half from your mother and the other half from your father.

What Is Outside of the Cells?

Mostly Water with Dissolved Elements and Compounds

Most of this water is in the tissue around the cells. The rest is inside the blood vessels that run through your whole body. Water makes up the fluid part of your blood called *plasma*.

What Is Blood Made Of?
Plasma and Blood Cells

RED BLOOD CELLS

Plasma is mostly water. It also contains some proteins, simple elements, sugars, and waste products from the cells. Blood cells travel through the body in plasma.

Red blood cells are very tiny, but there are trillions of them in your body. Their main job is to pick up oxygen from the lungs and take it to every cell in the body. The cells need oxygen to burn fuel. As the oxygen is used, carbon dioxide is formed. The red blood cells bring this carbon dioxide back to the lungs where it can be exhaled.

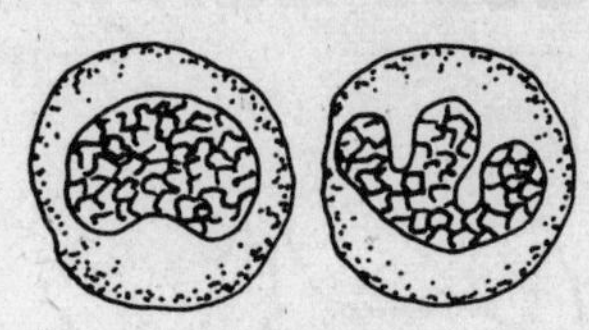
WHITE CELLS

In blood there are also several different kinds of *white blood cells*. They protect the body from harmful organisms and other particles. They do this by flowing around the organism and digesting it. There are about twenty billion of these helpful cells in the blood.

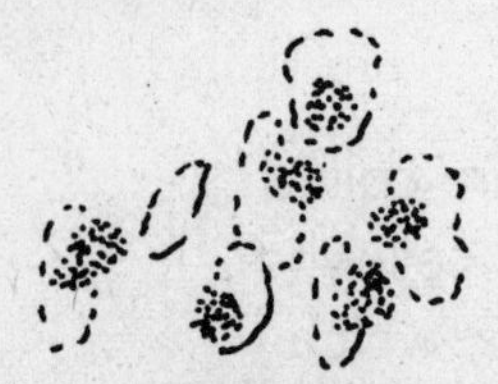
PLATELETS

Platelets are another type of blood cell with a very important job. Without platelets, cuts would not stop bleeding. When a blood vessel is damaged, platelets near the injury become sticky. They form a "plug" that slows the flow of blood. They also help your blood to clot.

DISCOVERY CORNER: Here is an example of how platelets help to stop a cut from bleeding. You will need a plastic kitchen funnel, 1/2 cup of raisins, and 1 cup of water. First soak the raisins in warm water until they are soft. The funnel represents a blood vessel that has been cut. Water pours through easily. Now pack the raisins lightly in the funnel. This forms a sticky plug. Little, if any, water will pour through. In the blood vessel, after the plug has formed, the blood begins to thicken or *clot*, and the opening is sealed.

What Do Blood Cells Come From?

Mostly from Bone Marrow

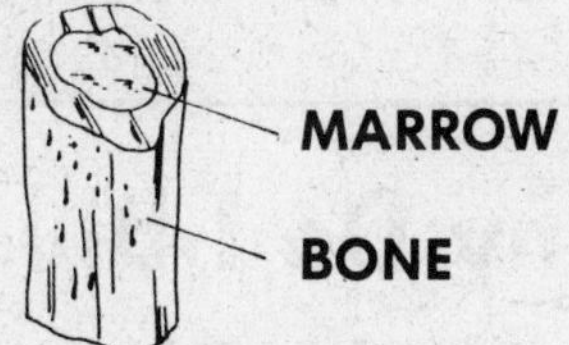

Inside of bones is a soft, spongy material called *marrow*. Platelets are produced in bone marrow. Red blood cells are also formed in the marrow, particularly in the bones of the chest, of the base of the skull, and of the upper arms and legs. Some white cells are produced in the marrow, but others develop in body organs such as the *liver* and *spleen*.

FUN FACT: Every day, about two hundred million red blood cells in your body wear out and are replaced.

How Does Blood Stay Clean?

By Passing through the Kidneys

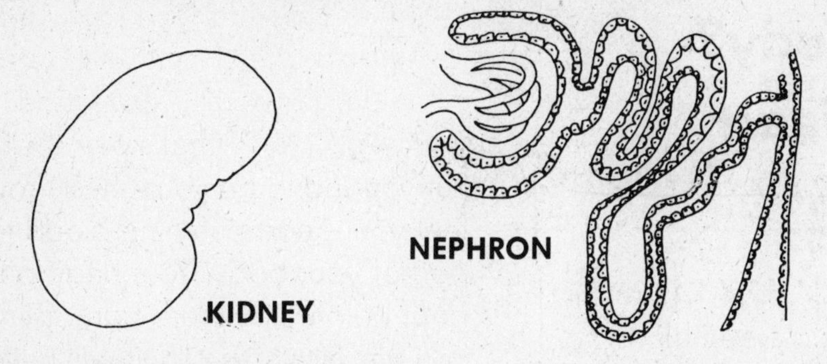

You have 2 *kidneys*. Each one is made of about a million tiny units called *nephrons* which are like fine strainers attached to long tubes. As blood flows through the kidneys, some of the plasma is strained off into the long tubes of the nephrons. The rest of the plasma continues on its journey through the blood vessels, along with the blood cells. From the nephron tubes, useful materials and water are carefully selected and returned to the bloodstream. Everything else is sent to the *bladder* where it is removed as urine. Using this process, the kidneys completely clean your blood about 60 times a day!

FUN FACT: When you are frightened, your blood will clot faster.

Why Do I Blush?

Because Blood Vessels near the Skin Surface Relax and Fill with Blood

When you are warm, embarrassed, or angry, the blood vessels near the skin surface relax and fill with blood. As a result, your skin turns pink.

FUN FACT: When you blush, your stomach lining reddens also.

Which Is the Hardest Working Part of My Body?

The Heart

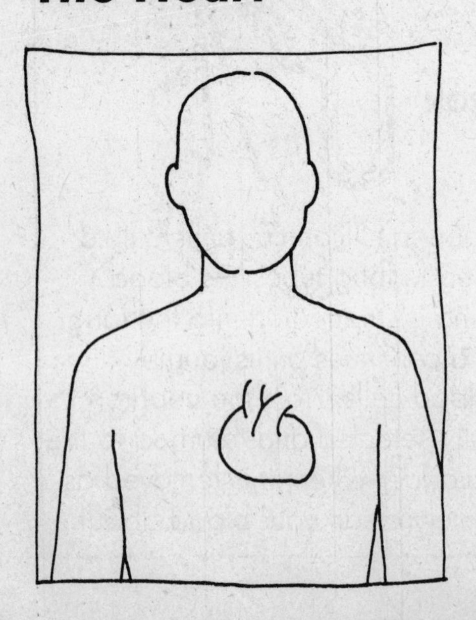

Put your hand on your chest. You can feel your heart pumping blood to every part of your body. Your heart began to beat before you were born. By the time you are 50 years old, it will have pumped more than 320,000 tons of blood!

FUN FACT: The average adult heart weighs about 12 ounces.

How Many Times Does My Heart Beat a Minute?

The Average Is About Seventy

You can actually count the waves of blood moving through your body with each beat of your heart by taking your pulse. Place your fingers gently on your throat until you can feel your heartbeat. If you have a watch or a clock with a second hand, you can measure your heart rate by counting how many beats you feel in 20 seconds. Multiply that by 3 and you will have the number of times your heart beats per minute.

When you are playing or working hard, your heart beats faster. Do 15 jumping jacks and take your heart rate again. What happened? What do you think will happen to your heart rate while you are sleeping?

DISCOVERY CORNER:
Doctors listen to your heart through a *stethoscope*. You can make a kind of stethoscope. All you need is the cardboard tube from an empty paper towel roll, and a friend. Now put one end gently at the center of your "patient's" chest and listen through the other end to the sound of a heart at work.

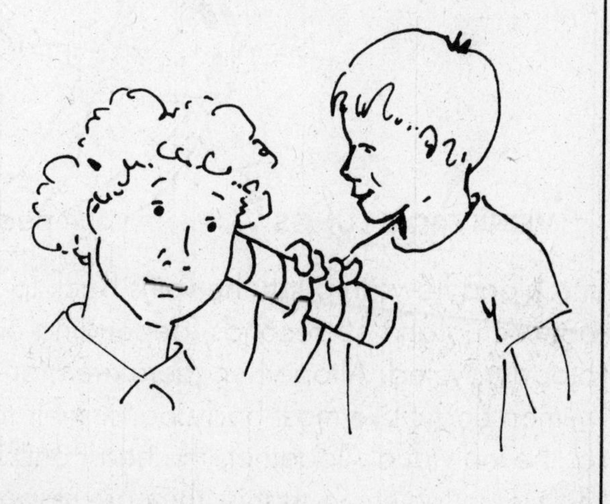

FUN FACT: The heart usually pumps about 5 quarts of blood per minute. That can rise to 20 quarts per minute during heavy exercise.

How Does Blood Get from My Heart to My Feet?

Through Arteries, Veins and Capillaries

Each beat of the heart sends oxygen-rich blood into the *arteries*. The arteries are flexible tubes, or blood vessels, that carry blood away from the heart to the rest of the body. As the arteries get further from the heart, they become smaller and smaller until they join the smallest vessels, called *capillaries*. In the capillaries, the blood cells drop off their cargo of oxygen and pick up waste products such as carbon dioxide. From the capillaries,

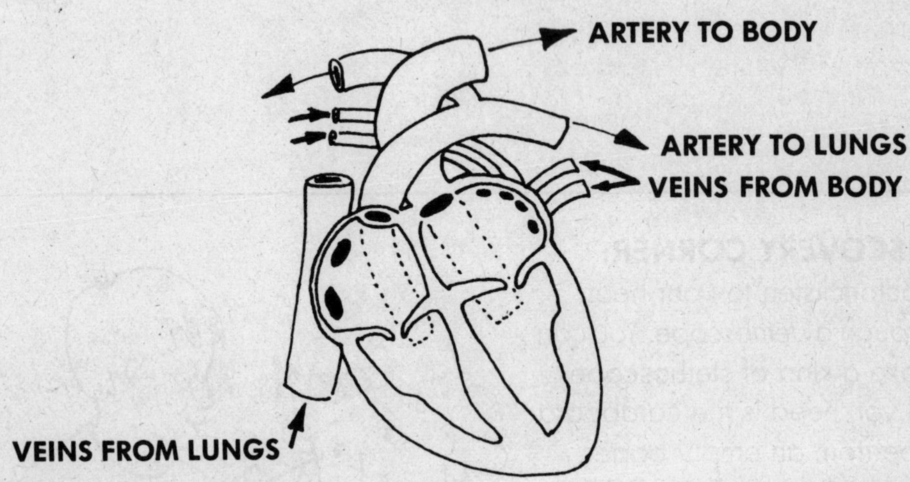

the blood flows through the *veins* back to the heart. The whole trip takes between 5 and 15 seconds, depending on the part of the body to which the blood traveled. Altogether, there are many miles of blood vessels in every human body. Like most body parts, their total length depends on the size of the individual. Together, the heart and blood vessels make up the *circulatory system* — that is, they are responsible for circulating blood through the body.

FUN FACT: If all of the veins, arteries, and capillaries of the average circulatory system were stretched out end to end, they would circle the Earth more than twice. That's more than 50,000 miles!

What Is a Bruise?

Blood under the Skin

There are many capillaries (small blood vessels) near the surface of the skin. Sometimes when you fall or bump into something, the capillaries can tear — even if you don't cut the skin. They repair themselves, but some blood can escape and flow under the surface of the skin, forming a bruise. Although a bruise may be a bit sore for a while, it is reabsorbed quickly by the body without a trace left behind. The next time you get a bruise, see how long it takes before it disappears completely.

How Does Air Get into My Body?

Through Your Nose and Mouth When You Inhale

Your nose and mouth are the beginning of an air passageway, or *respiratory system*, that goes all the way to your lungs. Lungs are like balloons filled with tubes, blood vessels, and millions of tiny air sacs called *alveoli*. Your breathing is controlled by movements of your chest and a large muscle below the lungs called the *diaphragm*. You breathe in when your chest expands and the diaphragm moves downward. This changes the air pressure in your chest. Your lungs expand and suck in air.

DISCOVERY CORNER: Here is how to build a model that shows how change in air pressure causes the lungs to expand. You need a pint jar with a tight-fitting lid, 2 plastic straws, a balloon, tape, and some modeling clay. Have a grown-up punch 2 holes in the lid, just big enough for the straws to fit through. Put the end of one of the straws into the balloon, and tape it completely. Then put both straws through the holes in the lid. Place the lid tightly on the jar.

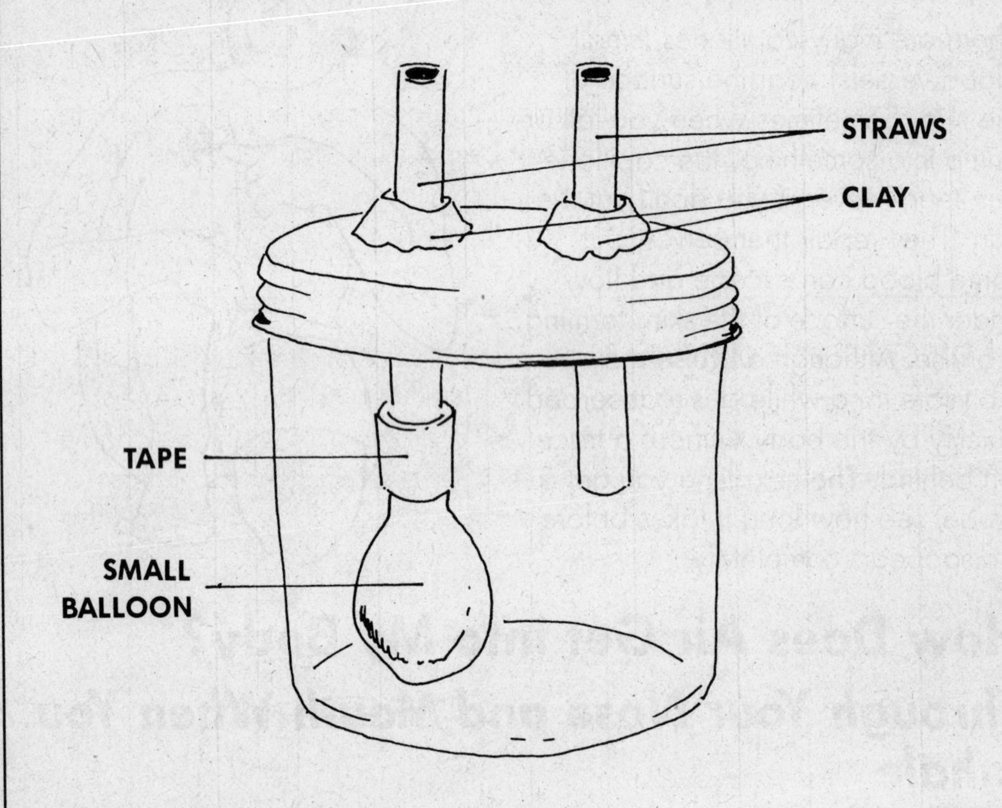

Pack modeling clay around the straws and near the holes in the lid. It is important that this seal is airtight. You lower the air pressure in the bottle by sucking on the straw without the balloon. What happens to the balloon? Let the air back into the jar and blow into the straw without the balloon. Now what happens to the balloon?

FUN FACT: No one knows for sure what causes hiccups. There is a man in the United States who has had a case of hiccups for over 50 years!

Why Does the Body Need Oxygen?

Because the Cells Use It To Release Energy

Food is "burned" in a sort of chemical fire to release energy to run the body. Without oxygen this would not be possible. Your body needs oxygen to burn the food.

DISCOVERY CORNER: To demonstrate the importance of oxygen, you will need modeling clay to hold a small birthday candle that will represent the fuel from food. Ask a grown-up to light the candle for you. The candle can burn because it is receiving oxygen from the air. Now, put a large glass jar over the candle so that it cannot receive oxygen. What happens to the flame?

How Much Air Is in Each Breath?

That Differs with Each Person

The amount of air you need changes depending on what you are doing. It also depends on the individual. People of different ages and sizes breathe in different amounts of air.

DISCOVERY CORNER: You can measure how much air is in each breath by building a *spirometer.* You will need 3 feet of one-quarter-inch rubber tubing, a plastic funnel, a one-gallon glass jug, plastic straws, a felt-tip pen, a measuring cup, and a large tub or aquarium. First you must mark the jug. With a measuring cup, fill the jug with water, one cup at a time. After you add each cup, make a line at the water level with a felt-tip pen. When the jug is full, pour out the water and label the marks like the bottle in the picture. Next, fill the tub or aquarium about 3/4 of the way with water. In the funnel, make a hole large enough to insert one end of the tubing, and place it in the water. Insert a straw into the other end of the tubing. This gives you a changeable mouthpiece. Fill your jug completely with water. Holding your fingers over the opening, turn the jug upside down and set it on the funnel. Now your spirometer is ready. Take a deep breath, exhale into the open end of the tube and count how many cups of air go into the jar.

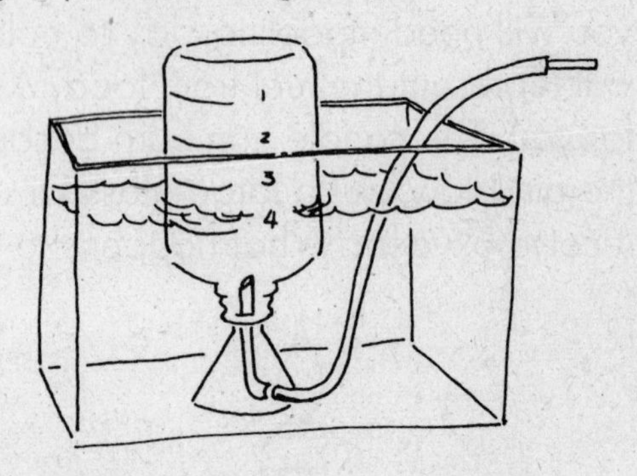

How Does Oxygen Get into My Blood?

It Is Exchanged in the Alveoli

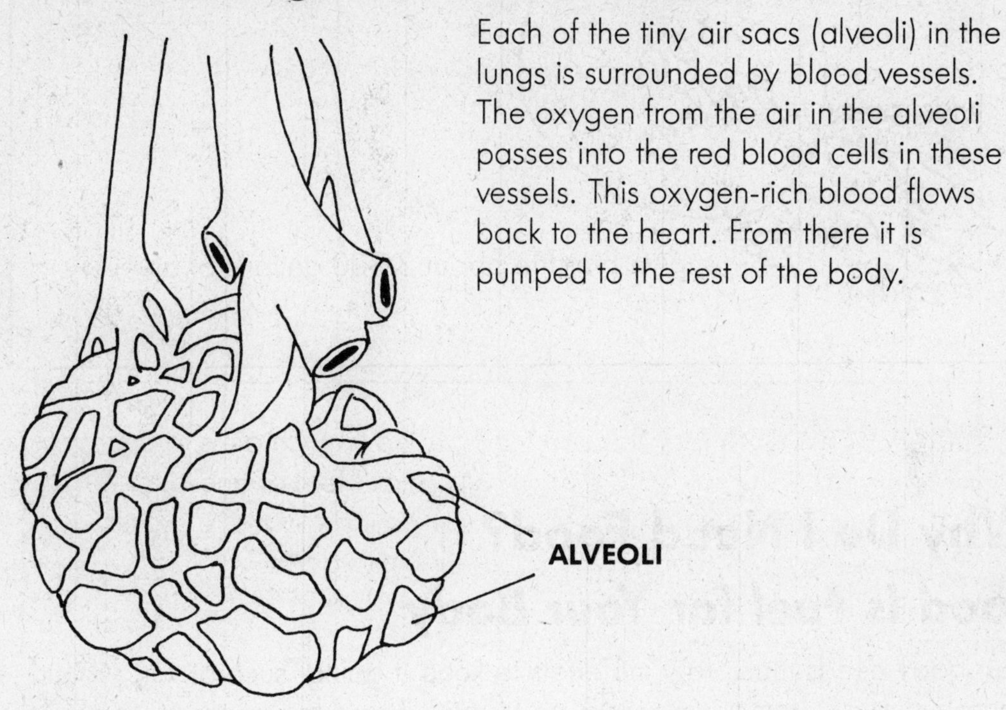

Each of the tiny air sacs (alveoli) in the lungs is surrounded by blood vessels. The oxygen from the air in the alveoli passes into the red blood cells in these vessels. This oxygen-rich blood flows back to the heart. From there it is pumped to the rest of the body.

Why Do I Have To Exhale?

To Get Rid of Carbon Dioxide

When blood cells arrive in the lungs, they drop off carbon dioxide and pick up fresh oxygen. The carbon dioxide is forced from the lungs when you exhale. You must get rid of carbon dioxide to prevent it from building up in the blood.

FUN FACT: A sneeze can reach the speed of a hurricane, nearly 100 miles per hour!

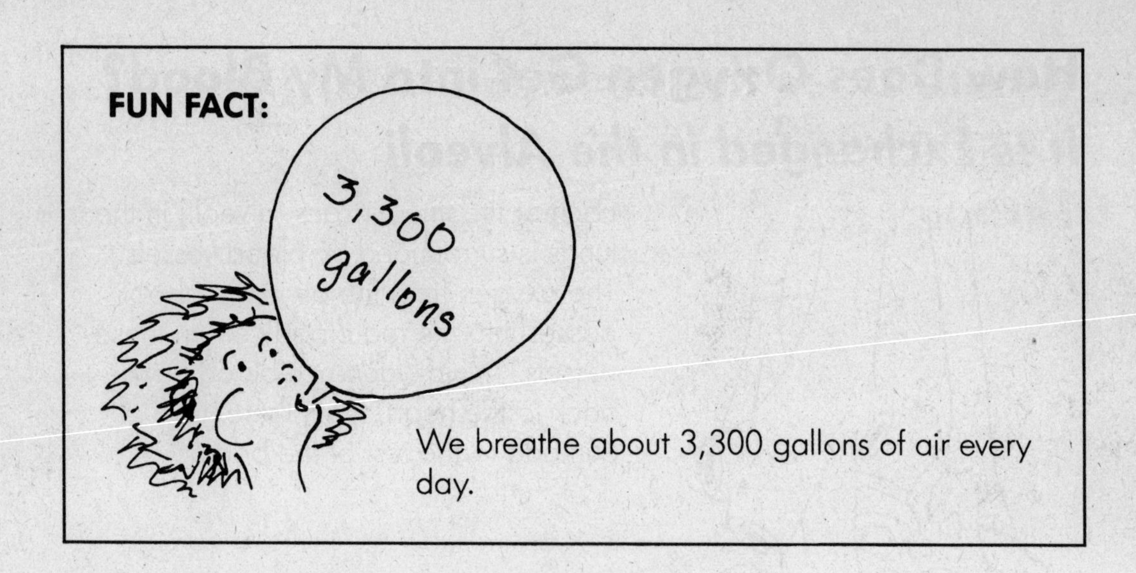

FUN FACT:

We breathe about 3,300 gallons of air every day.

Why Do I Need Food?

Food Is Fuel for Your Body

Your body needs many raw materials to keep it going, such as fat, protein, carbohydrates, vitamins, minerals, and water. These things are all in the food you eat. For the body to use these foods, it must break down the food into smaller compounds that can be absorbed. Protein and carbohydrates are easy for the body to break down because they dissolve easily in water. Digestion of fat is more complicated and requires a mixture from the liver called *bile*. Proteins, carbohydrates, and fats all can be burned by the body to release energy, or stored for later use.

FUN FACT: Food is moved down the throat by muscles, not gravity. It is possible to swallow even while standing on your head!

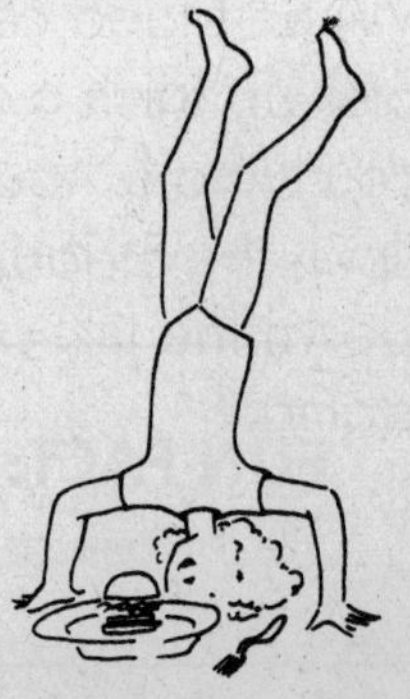

Where Does the Food Go after It Is Swallowed?

Through the Alimentary Canal

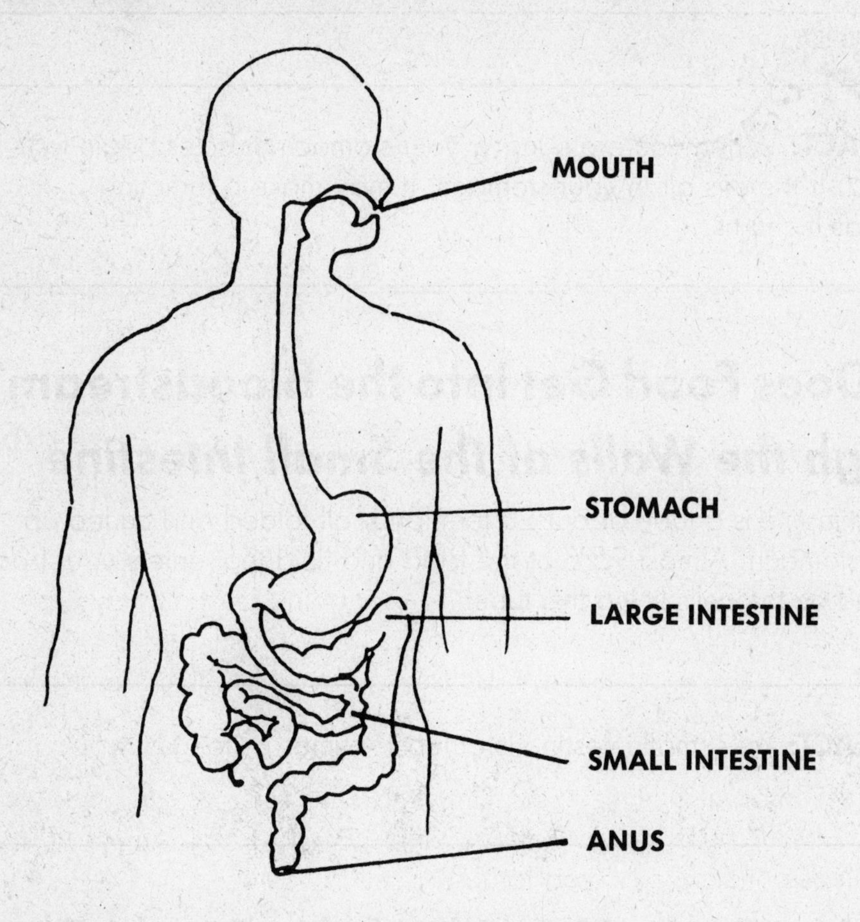

This canal is a tube that begins at the *mouth* and ends at the *anus*. The process of digestion begins in the mouth, where *saliva* mixes with, softens, and begins breaking down the food. After the food is swallowed, it travels down the first part of the canal called the *esophagus*, and arrives at the *stomach*, which is like a flexible bag. This trip takes about 10 seconds. In the stomach, the food is mixed with digestive juices and slowly released into the *small intestine*, where it will continue to be digested and then absorbed into the bloodstream.

DISCOVERY CORNER: Put a small piece of cookie or cracker in your mouth, but don't chew it. After only a few moments, the food will be soft enough to swallow even without chewing.

FUN FACT: When you are hungry, your stomach muscles begin to contract. If there is air in your stomach, it may make a rumbling sound as it churns.

How Does Food Get into the Bloodstream?

Through the Walls of the Small Intestine

The *small intestine* is a tube about 20 feet long, all folded and curled up under the stomach. Almost 95% of the food and fluid that enters your body is absorbed by the cells lining this tube.

FUN FACT: Your small intestine is at least 4 times as long as you are tall.

Does My Body Use All of the Food I Eat?

No

The material that is not absorbed by the small intestine is passed to the *large intestine*. It is called the large intestine because it is bigger around than the small intestine, even though the small intestine is longer. The large intestine absorbs mostly salt and water. The rest of the material in the large intestine travels to the *rectum*, where it will be stored and then eliminated.

DISCOVERY CORNER: Try this experiment to see how water is absorbed. Materials needed are a plastic soda bottle which will represent a part of the large intestine, 24 marbles and a teaspoon of salt mixed with a cup of water which will stand for material coming from the small intestine. You will also need a hand towel to represent the wall of the large intestine. Have a grown-up

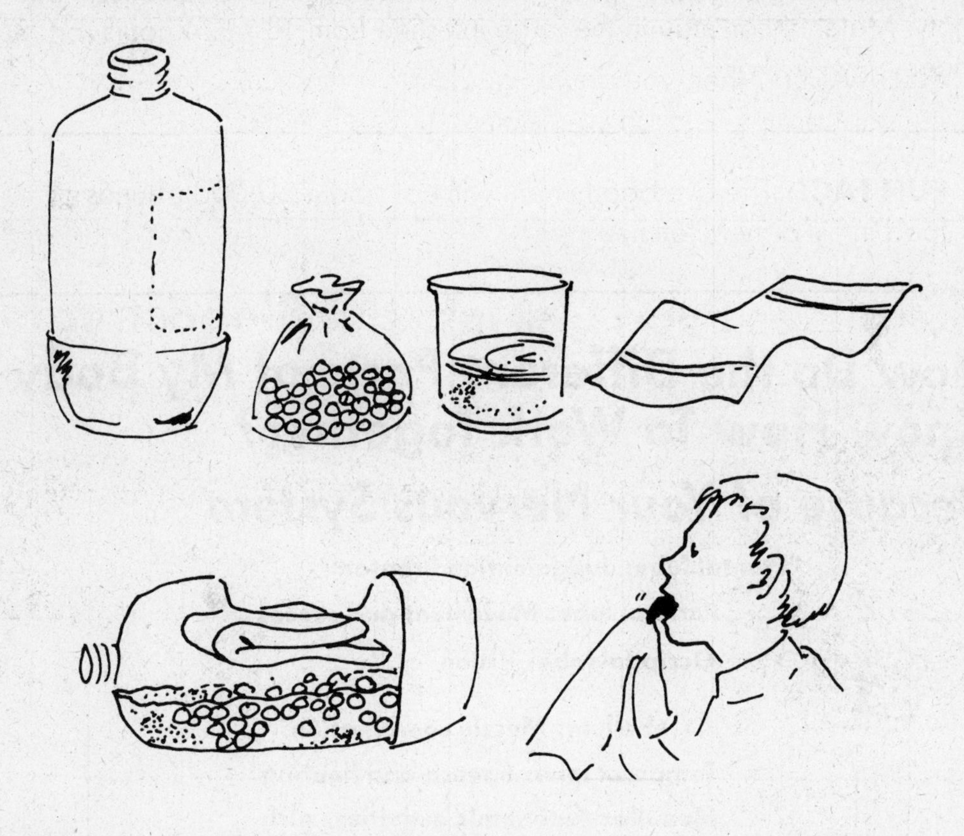

cut away about 1/3 of the side of the bottle so that you can reach the inside, but leave the pour spout intact. Lay the bottle on its side with the opening up, and pour the water and marbles inside. Fold the hand towel and tuck it into the opening. Let this set for about an hour, then remove the towel and pour the contents of the bottle into a bowl. The lining absorbed the water. Touch the towel to your tongue. Did the lining absorb the salt, too?

How Long Does it Take for a Meal To Travel through the Body?

On the Average, About Thirty-Four Hours

There are a lot of things that can affect this. Usually it takes from 3 to 6 hours for the stomach to empty and 3 or 4 hours for the small intestine to empty. Material can stay in the large intestine from 18 to 24 hours, or longer.

> **FUN FACT:** The average human will eat about 60,000 pounds of food in his or her lifetime.

How Do the Different Parts of My Body Know How To Work Together?

Because of Your Nervous System

Your body performs hundreds of tasks every day. Some of them, such as breathing and digestion, are done without thinking. These are called *involuntary* or *autonomic* activities. Things such as speaking and jumping rope are *voluntary* because you must think about doing them. Both kinds of activities are controlled by the *nervous system*. The center of this system is the *brain* and the *spinal cord*.

FUN FACT: The human brain is about 80% water.

DISCOVERY CORNER: The brain can coordinate many different activities at once. Look at the 3 circles below and touch one of them and say its number out loud. Following the above instructions involved thousands of nerves and several different areas of your brain. You had to read and understand the instructions, see the circles, choose one, move your hand, and coordinate the muscles of your throat, mouth, and tongue to speak. Of course, while you were thinking about this, you were also breathing, your heart was beating, and all of your normal body functions continued.

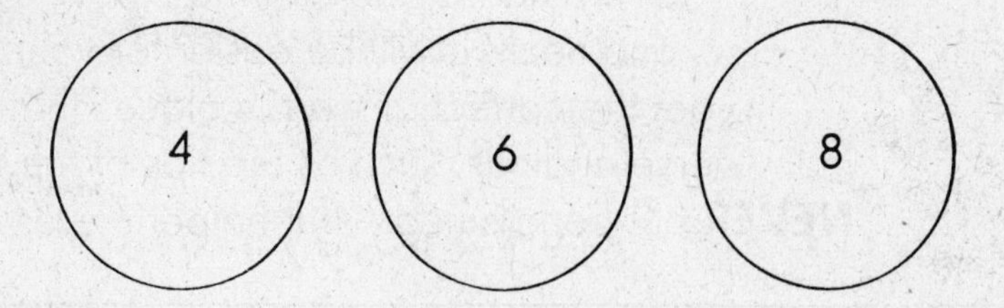

How Does the Brain Communicate with the Rest of the Body?

It Receives and Sends Signals through the Nerves

Nerves are collections of many individual cells (called *neurons*). Nerves are similar in function to telephone wires that carry messages. They are arranged in 43 pairs. Twelve pairs send signals directly to and from the brain. The other pairs carry messages to and from the brain by way of the spinal cord. The spinal cord is about as big around as your little finger. It is inside a hollow chain of bones called the *spinal column* or *backbone.*

FUN FACT: If placed end to end, the main nerves of the body would stretch about 45 miles.

DISCOVERY CORNER: Some parts of the body have more nerve endings than others. To test this you will need a helper and 2 pencils. Have your helper close his or her eyes and stretch out both hands, one palm up and one palm down. Now, holding the pencils side by side, gently touch one or both points on your helper's hand and ask how many points are touching. Try this in several areas of the hands, then try the elbow, shoulder, and neck. It will be easier for your partner to get the correct answer in places that have many nerve endings, such as the tips of the fingers. **NEVER** put pencils near your helper's eyes or ears.

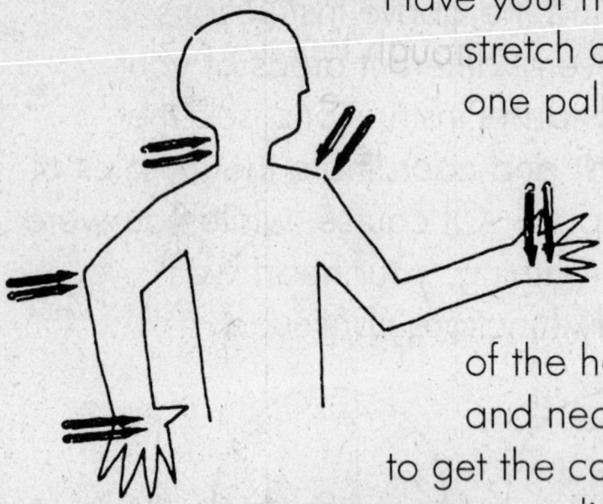

What Are Nerves Made Of?

Bundles of Neurons

Neurons are cells that have a cell body, a long tail called the *axon*, and branches at both ends. Some neurons are small and others very long, such as those which stretch from the toe to the spinal cord. Even though neurons don't actually touch each other, they communicate and form networks called *pathways*, and each pathway has a special job.

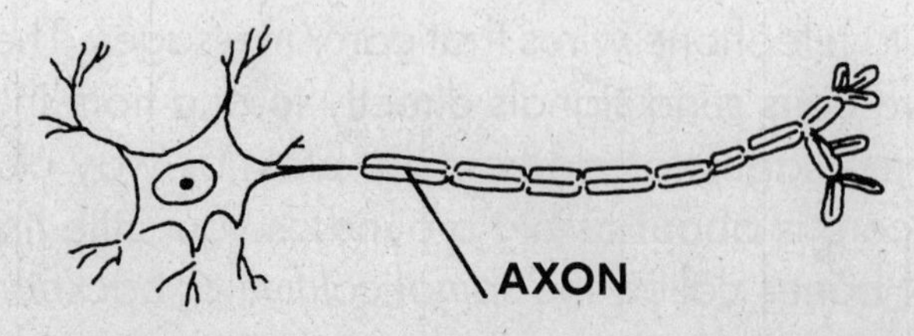

How Fast Does a Signal Travel to the Brain?

Some Can Travel as Fast as Two Hundred Fifty Miles Per Hour

Nerve signals can be acted on even before they reach the brain. If something flies toward your face, you will squeeze your eyes shut. Or if you touch something hot or sharp, you will quickly pull your hand away. That is because the signal takes a short cut through the spinal cord called a *reflex arc*. The brain feels the signal as pain, but the spinal cord has already sent an automatic command to your muscles to react.

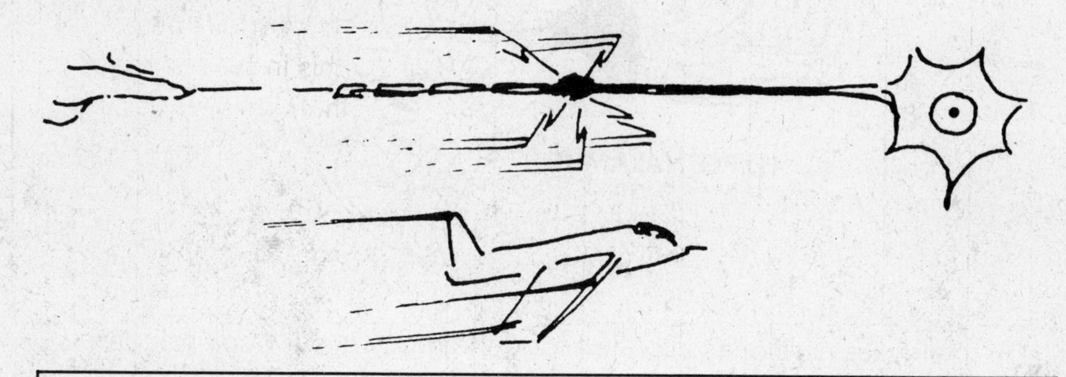

FUN FACT: Although certain parts of the brain can be less active while your body sleeps, the brain is always working.

Are Nerves the Only Messengers of the Body?

No

Many cells can be controlled by changing the amount of certain chemical messengers called *hormones*. Hormones help to regulate many functions of the body, such as growth and healing. They are produced in parts of the body called *endocrine glands* and released into the bloodstream.

Why Do I Feel Thirsty?

Because Your Body's Water Level Is Too Low

The *hypothalamus* is a very important gland in the brain. It has many jobs. When your body needs water your hypothalamus senses this and reacts in 1 of 2 ways.

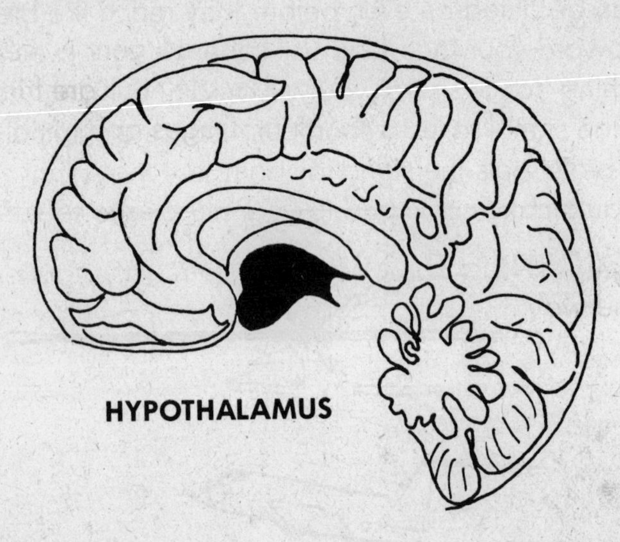

1) It releases hormones into the blood that cause your body to conserve the water you already have.

2) It sends a signal to the brain that will make you feel thirsty. This makes you want to drink.

The hypothalamus also controls another very important gland — the *pituitary gland*.

Why Do I Grow Taller?

Because of Your Pituitary Gland

This gland produces a growth hormone, which causes your body to grow. As your bones get longer, you get taller. Your maximum height is determined by your genes.

Why Does My Heart Beat Faster When I'm Afraid?

One Reason Is Adrenaline

Adrenaline is a hormone released from the *adrenal glands* when something frightens you or when there is an emergency. Many parts of the body react to adrenaline. The heart pumps faster. More blood gets to the muscles and less to the surface of the skin. That is why people who are frightened often look pale. You breathe faster, and the air passages open wider so more oxygen becomes available.

DISCOVERY CORNER: The breathing rate reacts to adrenaline much the same way it reacts to heavy exercise. You can make a chart to show how your breathing rate is affected. You need a watch with a second hand, paper, a pencil, and a ruler. Make a chart on the paper which looks like this . . . .

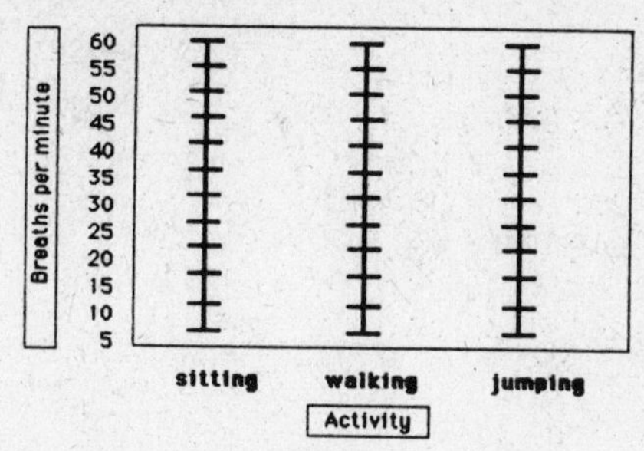

To take your breathing rate, sit quietly and count how many breaths you take in 15 seconds, then multiply this by 4. On your chart there is a line next to the word "Sitting." Using the ruler to measure a straight line, put a dot on this line next to the number of breaths you take per minute. Walk quickly around the room 10 times and take your breath rate again. Put a dot on the "Walking" line next to the new number. Now, jump up and down for 1 minute, take your breathing rate, and put a dot in its place on the chart above "Jumping." What is happening to your breathing rate? Are there other times when your breathing rate goes up?

Does Each Endocrine Gland Make One Hormone?

No

Some glands produce several hormones, and for many glands, making hormones isn't all they do. Here are a few of the important endocrine glands and the jobs of the hormones they produce.

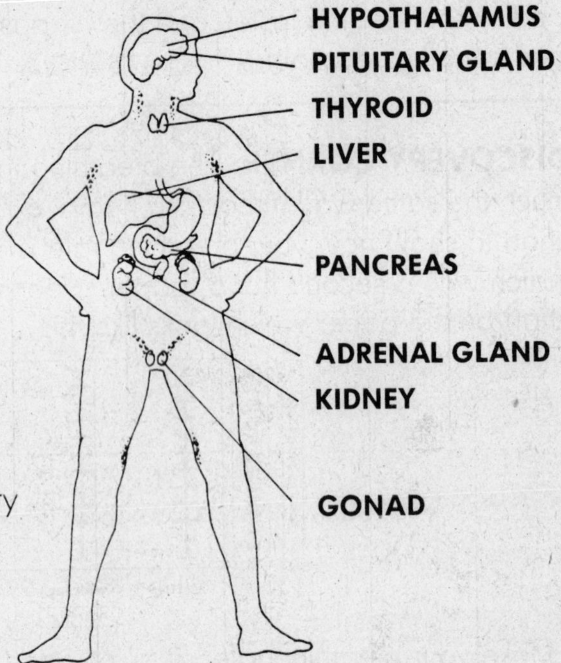

Adrenal gland:
Controls stress response
Reduces sex hormones
Regulates salt and water balance

Gonad:
Produces sex hormones

Hypothalamus:
Regulates water balance
Governs food intake
Controls activity of the pituitary gland

Kidney:
Controls red-blood-cell production
Regulates salt and water balance

Liver:
Regulates blood-sugar levels

Pancreas:
Regulates blood-sugar levels
Regulates production of digestive juices

Pituitary gland:
Governs the activities of other glands
Controls growth

Thyroid gland:
Influences temperature regulation
Governs energy production
Stimulates growth of the nervous system

Do All Glands Produce Hormones?
No

For example, *salivary glands* in the mouth and *sweat glands* in the skin do not. Glands that do not make hormones usually release special fluids onto the skin or into a body cavity.

What Do Sweat Glands Do?
They Help To Keep Your Body Cool

There are about two million sweat glands in the body of an average adult. Sweat glands are tiny tubes in the skin that help cool the body by releasing salty fluid called *sweat*. The sweat evaporates into the air and increases the cooling effect of the air.

DISCOVERY CORNER: To see how the evaporation of fluid can make things cool, try this experiment. You will need an old, long-sleeved shirt and a fan or air conditioner. Dip the left sleeve of the shirt in water and wring it out so that it does not drip. Put on the shirt and stand near the fan or air conditioner. Which arm is cooler, the one in the dry sleeve or the one in the wet sleeve?

You can also increase the cooling effect by exposing more skin to the air. Look at the two children lying on the beach. Which one is trying to get cooler?

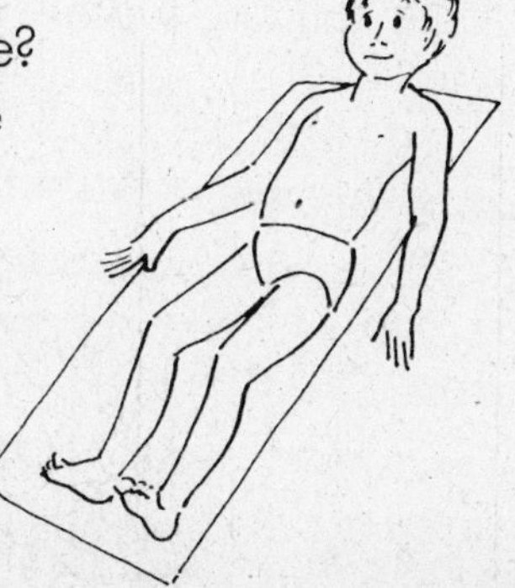

FUN FACT: There are more sweat glands in the palms of your hands and the soles of your feet than anywhere else in your body.

What Does the Skin Do?

It Protects the Body

An important job of the skin is to keep the inside of the body clean, moist, and protected. An average-sized adult has more than 9 1/2 square yards of skin. The skin is made up of 2 main layers. The inner layer is filled with sweat glands, blood vessels, hair roots, fat cells, oil-producing glands, and more. The outer level is a protective layer of dead skin cells that are constantly rubbed or washed away and are replaced from below.

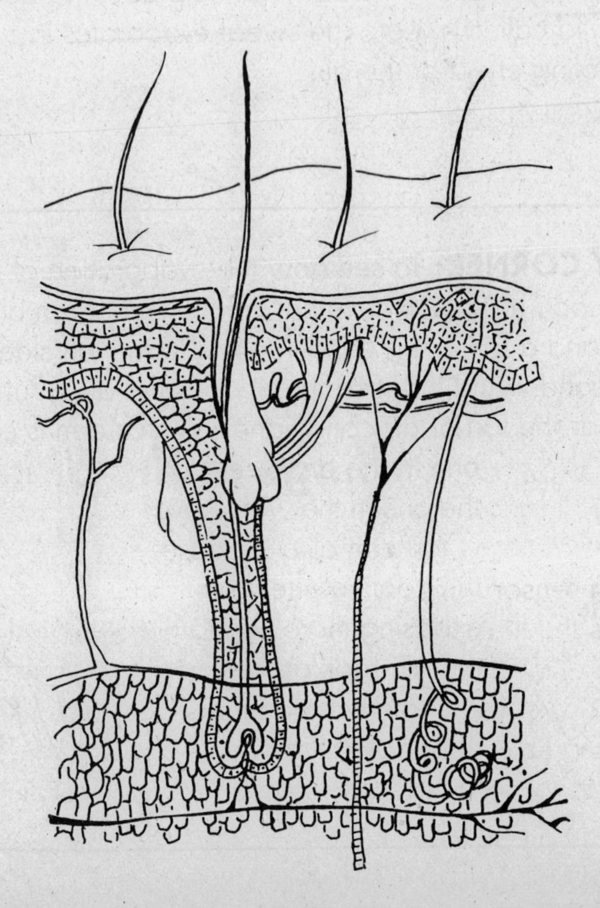

FUN FACT: Your skin completely replaces itself about once a month.

What Are Freckles?

Dots of Melanin

Melanin is coloring, or *pigment*, in the skin. It is there as protection against harmful rays from the sun. The more melanin there is, the darker your skin. Many people have tiny patches in the skin that contain lots of pigment. These are called freckles.

Why Do Cuts Hurt?

Because Nerve Endings in the Skin Are Irritated

Nerve endings in sensory organs collect information and send it to the brain. Your skin is the largest sensory organ. Other sensory organs are the eyes, ears, nose, and tongue. The skin sends your brain messages whenever you touch something. Touch includes 5 sensations: touch, pain, pressure, heat, and cold. There are more nerve endings to detect pain than any of the other sensations. The least amount of nerve endings is for heat and cold.

DISCOVERY CORNER: A bowl of water can feel hot and cold to you at the same time. To try this experiment you must have 3 mixing bowls. Fill one with cold water from the tap. Fill another with very warm, but not hot, water, and fill the third with water at room temperature. First, soak your left hand in the warm water, and your

right in cold, for 3 or 4 minutes. Now, put both hands at once into the room temperature water. It will feel cold to the left hand and warm to the right. Each hand senses the room temperature water in comparison to the water it was soaking in before, so it feels warmer to one and cooler to the other.

What Are the Five Senses?

Touch, Sight, Hearing, Taste, and Smell

Through the senses, your brain experiences the world around you. The skin, eyes, nose, ears, and tongue send information to the brain through nerve pathways.

Do Both Eyes See the Same Thing?

No

Your eyes are the organs of sight that send signals to the brain. Each eye sees an object from a slightly different view. Stare directly at something in front of you. Squeeze one eye shut, then quickly open it and close the other. The object will appear to move slightly back and forth. When both eyes are open, your brain combines the information from each, and you see one view of the object. It is important to have this double view because it helps to figure out distance, particularly for things that are close to you.

DISCOVERY CORNER: You can test this with 2 pencils. With your arms at shoulder height and about 2 feet apart, hold a pencil in each hand. Close one eye and try to touch the pencil points together. Now try it with the other eye closed. Try it again with both eyes open. Which way is easiest?

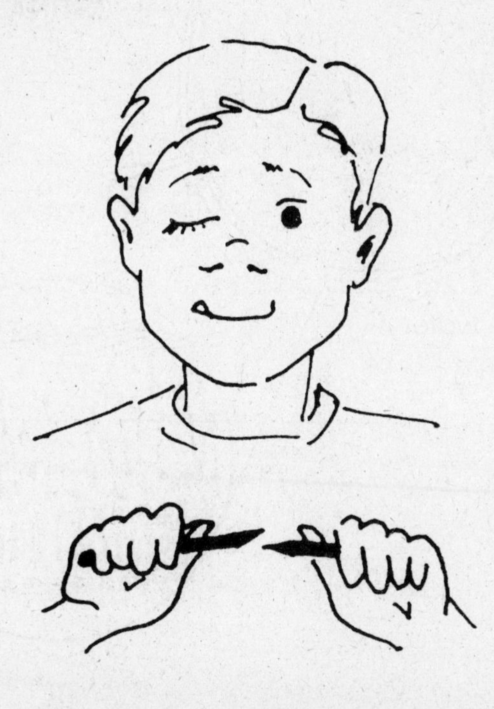

What Makes the Eyes Work?

The Effect of Light on Nerve Endings

All of the parts of the eye work together. The *cornea* is something like a window that lets in light. The *iris* is the colored part of the eye. It helps control the amount of light that enters. The *pupil* is the black opening in the center of the iris. It becomes smaller or larger depending on the amount of light reaching the eye. When it is darker outside the pupil becomes larger to let in more light. The *lens* is a tissue that adjusts to focus on things far away or close up. The *retina* is the lining of the inside of the eye. It has many nerve endings that react when light falls on them. The *optic nerve* is the pathway to the brain that carries nerve signals.

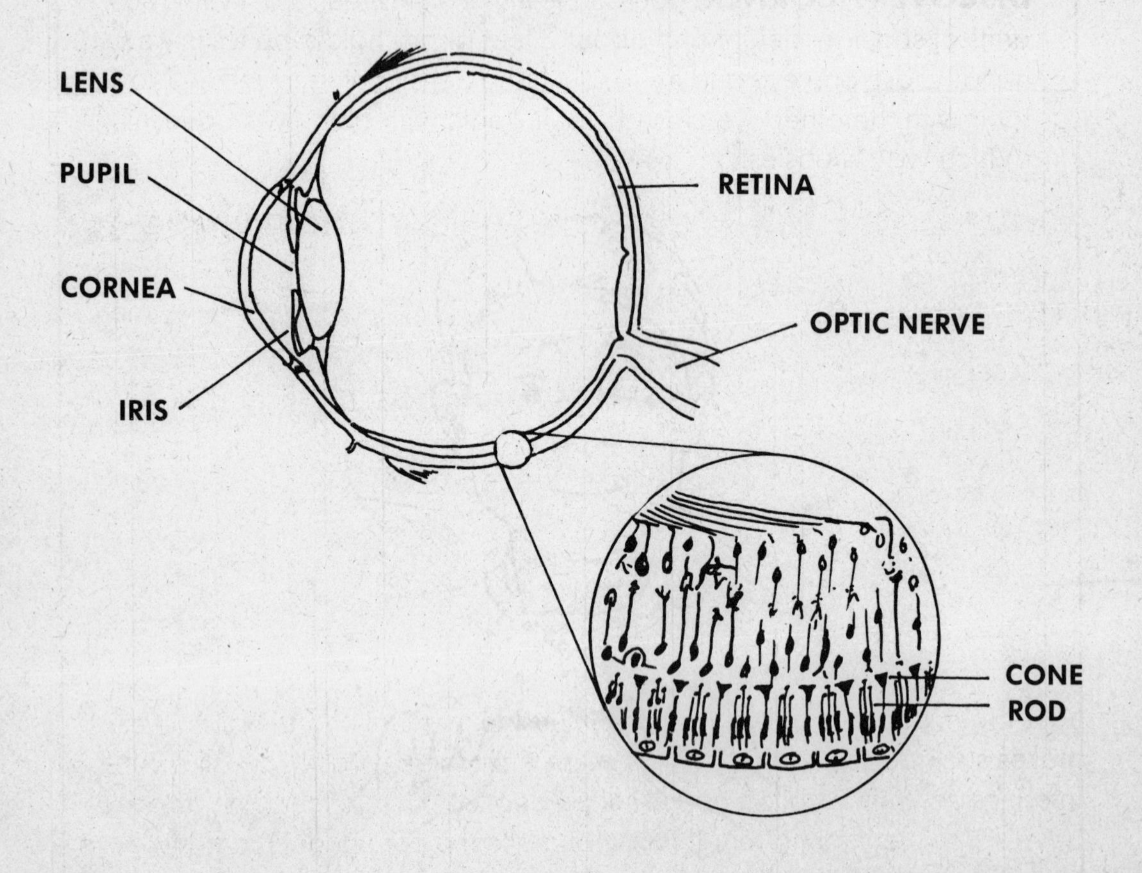

Why Do I See Color?

Because of Special Cells in the Retina

The retina or lining of the eye is filled with little cells called *rods* and *cones*, called such because of their shape. Cones are sensitive to color and need a lot of light to work effectively. Rods are not sensitive to color and can be activated by less light. Rods allow you to see at night, but objects are not as clear and appear in shades of gray.

FUN FACT: Certain animals probably do not see things in color as we do. Scientists think this is so because the animals' eyes have rods but no cones.

How Can I Tell How Far Away Something Is?

By Using Both Eyes

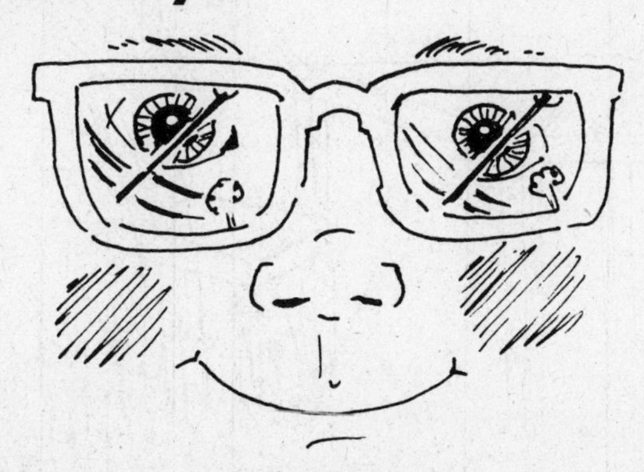

Your brain judges distance by combining the information from both eyes. It makes such decisions about the outside environment based on all available information. This is called our visual perception. When the brain does not have complete information, it may not make the correct decision. How we perceive an object and how that object really exists may be quite different.

DISCOVERY CORNER: Look at this table. Is it big or small?

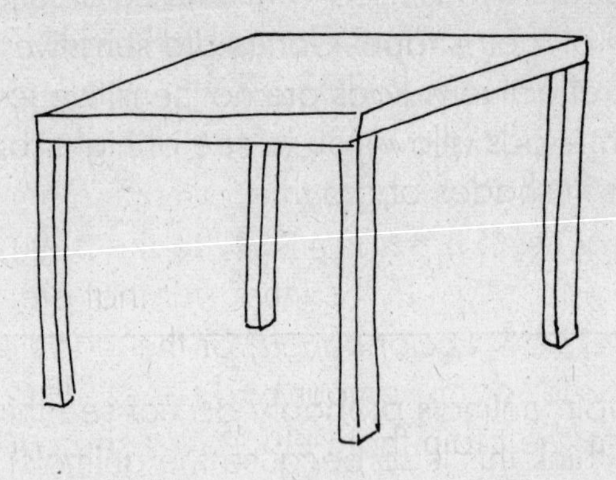

You don't have enough information yet to tell what size this table is. Now let's add some information to the picture. Look again.

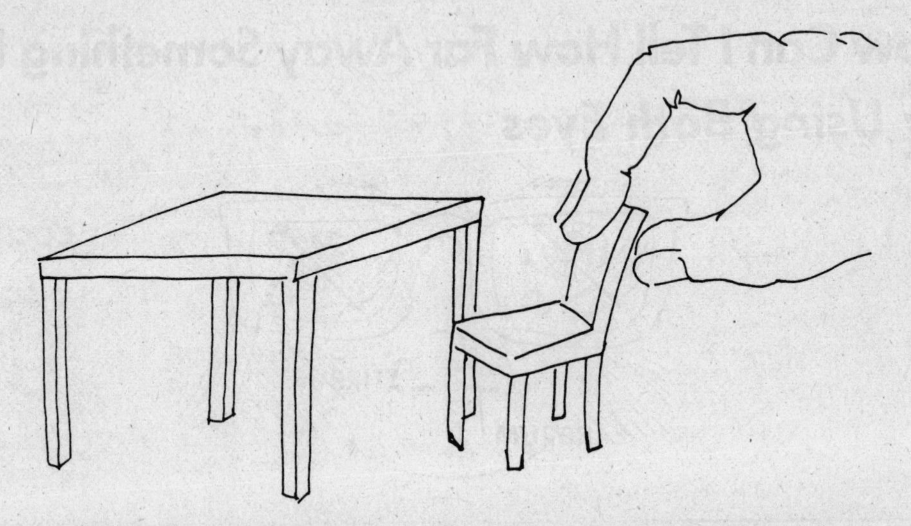

Now how big do you think the table is? The hand in the second picture gave your brain the information you needed to figure out that the table is a tiny one.

What's Inside My Ears?

The Middle and Inner Ear

Sound entering the *outer ear* is felt as pressure on the *eardrum* which causes the eardrum to vibrate. The vibration is passed on to the 3 tiny bones of the *middle ear*: the *hammer, anvil*, and *stirrup*. Next, the vibration travels to a fluid-filled tube lined with tiny, hairlike nerve cells in the *inner ear*. This tube is called the *cochlea*. From here, the information is passed to the brain. The *auditory nerve* is the pathway to the brain.

There are also special organs in the inner ear that are not used for hearing. These are the *semicircular canals*, or the *utricle* and *saccule*. When you turn or tip your head, or move your body, these little organs sense this and send messages to the brain that help you to keep your balance.

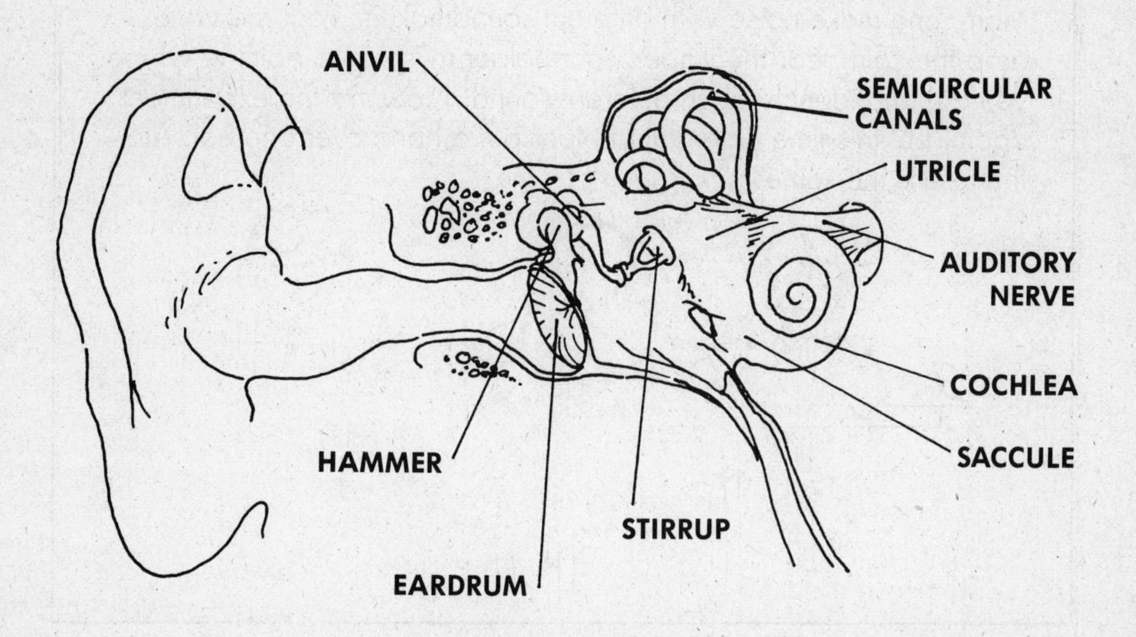

FUN FACT: Space is almost a vacuum with no air molecules to carry vibrations. You would not hear an explosion in outer space as you would on Earth.

Do Both Ears Hear the Same Thing?

No

There are small differences between the sounds that reach each ear. The ear nearest the sound source hears the sound a tiny bit sooner and a little louder than the ear farthest from the sound. The brain uses this different information to distinguish the direction from which a sound is coming.

DISCOVERY CORNER: Try this experiment with a partner to test his or her ability to distinguish sound and direction. You will need some ''soundmakers,'' such as a pitcher of water and a glass, a sheet of paper, and a coin. Blindfold your helper and have him or her stand in the center of the room. Move from place to place in the room, and make noise with different soundmakers. Pour the water, drop the coin, tear the paper. Your helper must try to point to where you are and identify each different sound. Now, try the experiment again, but this time have your helper put a hand over one ear. Are the results the same?

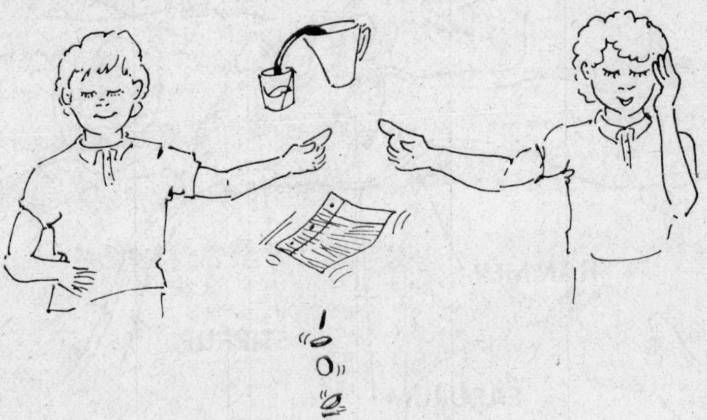

FUN FACT: It is not the sea you hear when you put a shell to your ear. It is the sound of the blood flow in your ear.

How Do I Taste Food?
With Taste Buds

There are about 10,000 of these tiny receptors in your mouth. Most of them are on your tongue, but there are many on the roof of the mouth and in the throat, too. Look at your tongue in the mirror. All those little bumps you see are *taste buds*. Taste buds on the tip of the tongue react to sweetness and saltiness, the ones at the back are sensitive to bitterness, and the ones on the side react to sour or salty tastes.

> **FUN FACT:** Everyone in the world has a unique set of fingerprints and a unique tongueprint, too.

Why Do Foods You Like Usually Smell Good?

Because Your Senses of Smell and Taste Often Work Together

When you have a bad cold and your nose is stuffy, food doesn't seem to have much flavor. Although you are still able to sense salty, bitter, sweet, and sour, it is hard to distinguish anything else without the sense of smell.

DISCOVERY CORNER: Try this experiment. You will need a peeled apple, a peeled potato, and a helper. Grate the apple into one dish and the potato into another. You may need a grown-up to help with this. Now, blindfold your helper and have him hold his nose, eat some of each food, and try to identify it. Try it yourself. Is it difficult to identify the food without smelling it?

How Does the Nose Detect an Odor?

With Tiny, Hairlike Cells

There are about twenty million nerve cells in your nose. These cells are bathed in a thick fluid called *mucus*. Molecules from the substance that you are smelling drift into the nose, mix with the mucus, and cause a reaction in the nerve cells. This information is then sent to the brain.

What Gives My Nose Its Shape?

Mostly Cartilage

Cartilage is a tough material that shapes your ears and nose. Cartilage also acts as padding between bones. It cushions the bones in your spine. Cartilage absorbs the bumps and shocks of walking and allows the backbone to remain flexible.

FUN FACT: Gravity pulls down on the cartilage in your spine. Astronauts who spend time in zero gravity come back to Earth taller than when they left.

What Are Bones Made Of?

Living Tissue and Calcium

Most bones have a soft inner core and a hard outer covering. The soft, spongy *marrow* inside your bones is living tissue with tiny blood vessels running through it. The marrow is where new red blood cells are manufactured. The hard outer covering of the bones that make up your skeleton contains about 99% of the calcium in your body. Your skeleton is what gives your body its shape.

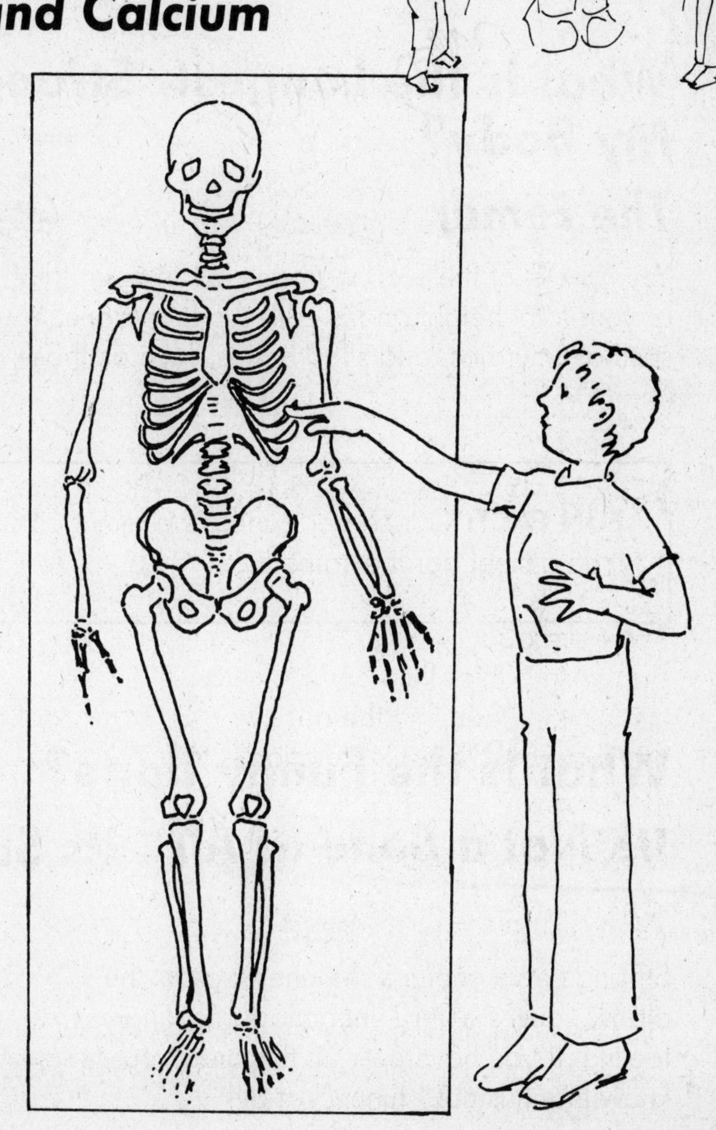

How Many Bones Do I Have?

About Two Hundred Six

Babies are born with around 350 bones, but many grow together as the baby becomes older. For instance, a baby's skull is made up of more than 20 different bones. As a child becomes an adult, these bones grow together into a solid cover to hold and protect the brain.

What Is the Longest, Strongest Bone in My Body?

The Femur

You have 2 of these strong bones, 1 in each thigh. The *femur* is about 1/4 of your total height and is called a *long bone*. You also have *flat bones*, such as your ribs, and *short bones*, such as those in your wrists and ankles.

FUN FACT: Ounce per ounce, the femur can support more weight than a steel bar the same size.

What Is the Funny Bone?

It's Not a Bone at All

Striking nerve endings stretched across the elbow causes a very uncomfortable, tingly feeling. If you have ever hit this area you know that it is not ''funny'' at all!

How Are Bones Held Together?
With Ligaments

Ligaments are like strong, flexible cords. They join bones across movable joints.

Is My Elbow a Bone?
No, It Is a Joint between Two Bones

You are able to bend certain parts of your body because of *joints*. You have different kinds of joints depending on what sort of movement is needed. The hip and shoulder joints are *ball-and-socket joints*. This kind of joint allows the legs and arms to swing back and forth or out to the sides.

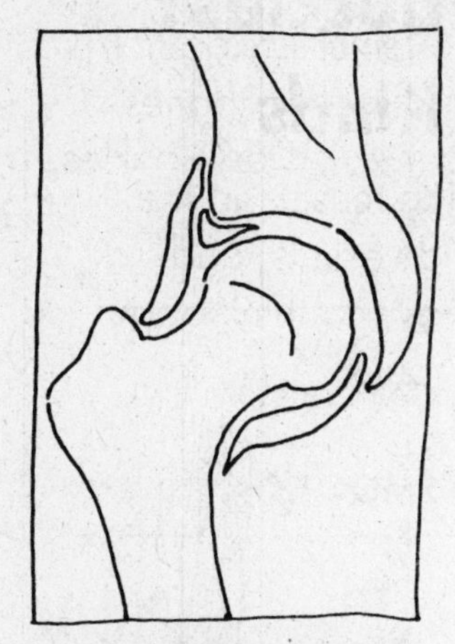

BALL AND SOCKET

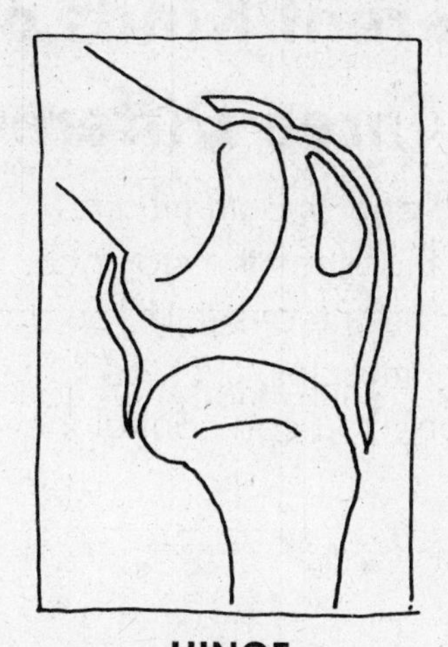

HINGE

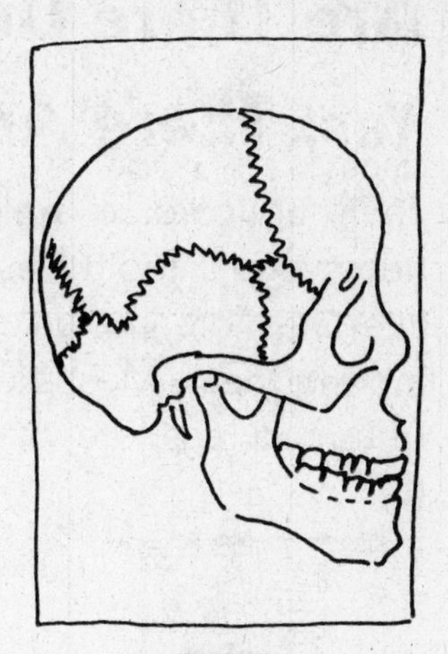

FIXED

Joints in the knees and elbows bend in only one direction. These are *hinge joints*. You also have *fixed joints*, such as those in the ribs and skull, which don't move at all.

FUN FACT: The sound made when you "crack" your knuckles is the sound of gas bubbles popping in your joints.

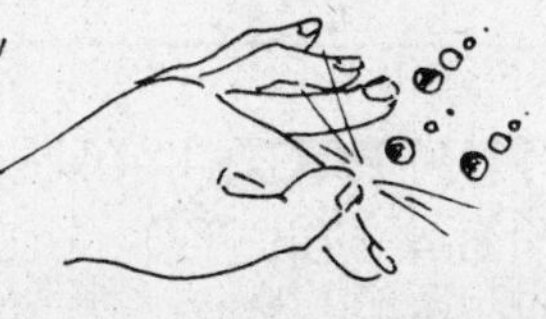

What Are Muscles?

Bundles of Cells

Muscles are bundles of cells called *muscle fibers*. The muscle fibers are all bound together by a strong connecting tissue filled with nerves and blood vessels. Small muscles can be made up of a few hundred fibers and large muscles can contain a few hundred thousand! You have about 650 muscles. Those that move bones are connected to them by a strong cord called a *tendon*.

Are There Different Kinds of Muscles?

Yes, There Are Three Different Kinds

There are *cardiac, skeletal*, and *smooth muscles*. Cardiac muscle in the heart, and smooth muscle that lines the stomach, blood vessels, and intestines, are involuntary. That means that you don't have to think about moving them. Most skeletal muscles, such as those in your arms and legs, are voluntary. You can control them.

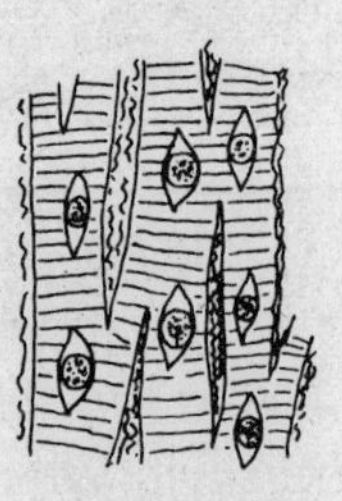

CARDIAC

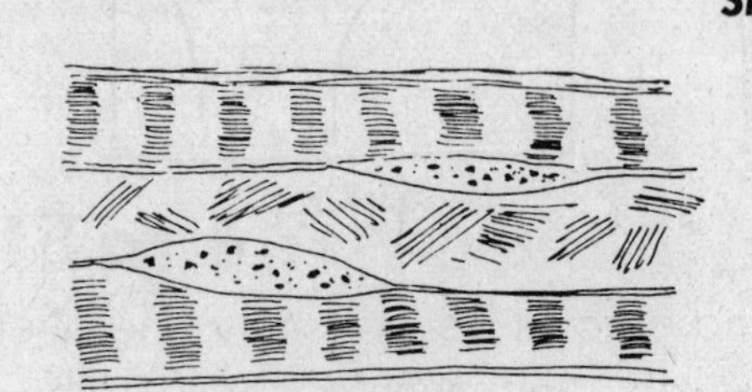

SKELETAL

SMOOTH

FUN FACT: It takes 17 muscles to smile, 43 muscles to frown, and more than 200 muscles to take one step.

How Do Muscles Move the Body?

By Working in Pairs

Skeletal muscles work in pairs because they can only pull, not push. When you bend your arm, your *biceps* and *triceps* muscles work together. When you bend your arm, the muscle fibers in your biceps shorten and pull on the tendon connected to one of the bones in your lower arm. This is something like workmen pulling a load with a rope. When you want to straighten your arm, the biceps relaxes and gets longer again, and the triceps shortens and pulls your arm the other way.

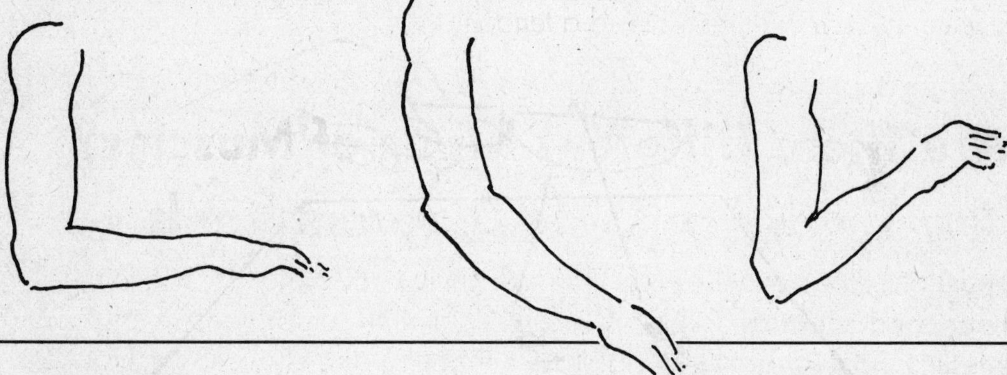

FUN FACT: By contracting and relaxing very quickly, muscles produce heat inside the body when you're cold. Sometimes there can be as many as 10 or 20 contractions per second. This is known as shivering.

Do I Use Muscles When I'm Standing Still?

Yes, Always

When you are standing still you are using many involuntary muscles that are involved in breathing and other things. Although you may not think about it, you are also using voluntary muscles to keep your balance and hold up your head. If a person sitting in a chair becomes sleepy, his head will begin to nod. If he actually falls asleep, there is a chance he could even slide out of the chair! This is because you are not in complete control of your voluntary muscles as you fall asleep.

DISCOVERY CORNER: The cells that make up a muscle take turns tightening and relaxing. So even if you try to hold your hand very still, the muscles will still move. To prove this you will need a piece of wire 3 inches long and a butter knife. Bend the wire into a "V" shape. Hook it over the edge of the butter knife, and rest the ends of the wire lightly on a table. Without resting your hand on anything, try to hold the wire still. Can you? You can't hold the wire still because the muscles you are using to hold the knife continue to tighten and relax, making your hand move slightly.

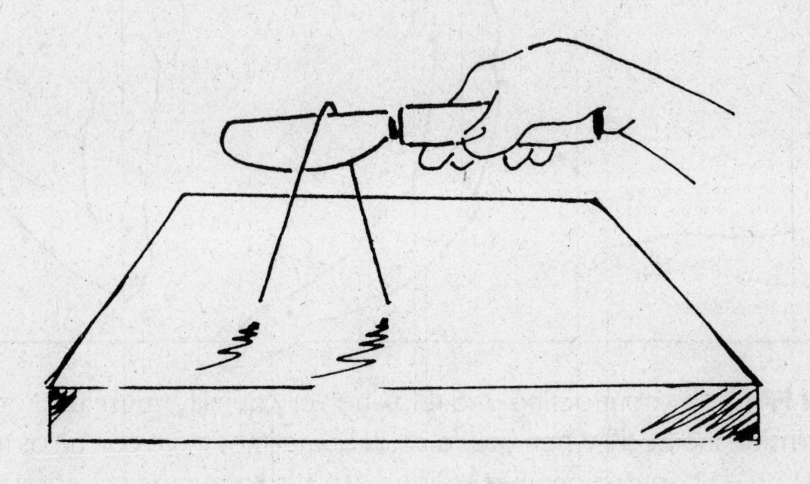

Perhaps there are many other things you would like to know about the human body. Remember, you have one of the best tools to find the answers, the remarkable human brain.